I0884507

21st-Century Engineering Solutions for Climate Change

EXTREME WEATHER EVENTS

JEANNE MARIE FORD

Cavendish Square

New York

Published in 2019 by Cavendish Square Publishing, LLC
243 5th Avenue, Suite 136, New York, NY 10016

Copyright © 2019 by Cavendish Square Publishing, LLC

First Edition

No part of this publication may be reproduced, stored in a retrieval system, or transmitted in any form or by any means—electronic, mechanical, photocopying, recording, or otherwise—without the prior permission of the copyright owner. Request for permission should be addressed to Permissions, Cavendish Square Publishing, 243 5th Avenue, Suite 136, New York, NY 10016. Tel (877) 980-4450; fax (877) 980-4454.

Website: cavendishsq.com

This publication represents the opinions and views of the author based on his or her personal experience, knowledge, and research. The information in this book serves as a general guide only. The author and publisher have used their best efforts in preparing this book and disclaim liability rising directly or indirectly from the use and application of this book.

All websites were available and accurate when this book was sent to press.

Library of Congress Cataloging-in-Publication Data

Names: Ford, Jeanne Marie, 1971- author.
Title: Extreme weather events / Jeanne Marie Ford.
Description: New York : Cavendish Square, 2019. | Series: 21st-century engineering solutions for climate change | Audience: Grades 7-12. | Includes bibliographical references and index.
Identifiers: LCCN 2017051184 (print) | LCCN 2017056750 (ebook) |
ISBN 9781502638403 (ebook) | ISBN 9781502638380 (library bound) | ISBN 9781502638397 (pbk.)
Subjects: LCSH: Climatic extremes. | Climatic changes. | Severe storms. | Global warming. | Weather.
Classification: LCC QC981.8.C53 (ebook) | LCC QC981.8.C53 F674 2019 (print) | DDC 363.34/92--dc23
LC record available at https://lccn.loc.gov/2017051184

Editorial Director: David McNamara
Editor: Kristen Susienka
Copy Editor: Rebecca Rohan
Associate Art Director: Amy Greenan
Designer: Alan Sliwinski/Megan Mette
Production Coordinator: Karol Szymczuk
Photo Research: J8 Media

The photographs in this book are used by permission and through the courtesy of: Cover Chip Somodevilla/Getty Images; p. 3 Sasa Kadrijevic/Shutterstock.com; p. 8 Designua/Shutterstock.com; p. 11 Eniko Balogh/Shutterstock.com; p. 16 Stephen J. Krasemann/Science Source; p. 22 ©AP Images; p. 25 Bloomberg/Getty Images; p. 27 New York Daily News/Getty Images; p. 28 Michael Neil Thomas/Shutterstock.com; p. 32 Marcus Yam/Los Angeles Times/Getty Images; p. 36 RicciPhotos/iStock/Thinkstock; p. 38 Justin Sullivan/Getty Images; p. 40 Marty Melville/AFP/Getty Images; p. 44 Steve Mack/Getty Images; p. 46 left Jitalia17/E+/Getty Images; p. 46 right Richard Whitcombe/Shutterstock.com; p. 50 The Washington Post/Getty Images; p. 54 Matej Hudovernik/Shutterstock.com; p. 57 Belinda Pretorius/Shutterstock.com; p. 58 Julie Dermansky/Corbis/Getty Images; p. 63 Jim Brandenburg/Minden Pictures/Getty Images.

Printed in the United States of America

CONTENTS

CHAPTER ONE

FACING EXTREME WEATHER

In August 2017, Hurricane Harvey made landfall over south Texas and dumped record amounts of rain over four days. Scientists called it a once-in-a-thousand-years weather event. Days later, Hurricane Irma chugged up the Atlantic coast. It cut a path of death and destruction across the Caribbean and the US state of Florida. Two weeks later, Hurricane

Opposite: Texas is vulnerable to hurricanes, such as the one pictured here.

Maria battered the island of Puerto Rico, knocking out power to all of its 3.4 million people and causing a massive humanitarian crisis.

If it seems that severe storms have become more frequent and intense in recent years, it's because they have. Nine of the ten costliest Atlantic hurricanes have occurred since the year 2000. Hurricane Harvey, which happened in 2017, is at the top of the list.

There are many reasons that hurricanes have been so costly and deadly in the twenty-first century. The world's population is growing, so more residents have been affected. Property values have increased. Large numbers of people have moved into low-lying areas that flood more often. Finally, the storms themselves have become more severe.

Hurricanes aren't the only type of extreme weather on the rise. As Harvey hit the United States, a record-breaking heat wave struck San Francisco, California. At the same time, intense wildfires devastated parts of the Pacific Northwest. Many people asked the same question: Why?

Weather Patterns

Severe weather has been a part of Earth's history since prehistoric times. The Bible and the *Epic of Gilgamesh* both tell the story of an ancient flood so cataclysmic that it nearly wiped out all living things. Earth has experienced at least five major freezing events, called ice ages, over the course of millions of years. Climate change is a natural part of the planet's evolution.

In modern times, the climate is changing more quickly than it ever has before. The average temperature has risen by about 2 degrees Fahrenheit (around 1 degree Celsius) since modern recordkeeping began in 1880. The year 2016 was the warmest on record. It was the third year in a row to reach this milestone.

While there will always be some temperature differences from year to year, this warming trend is expected to continue and intensify. Some scientists predict that the average temperature may increase by as much as 11°F (6°C) over the next century.

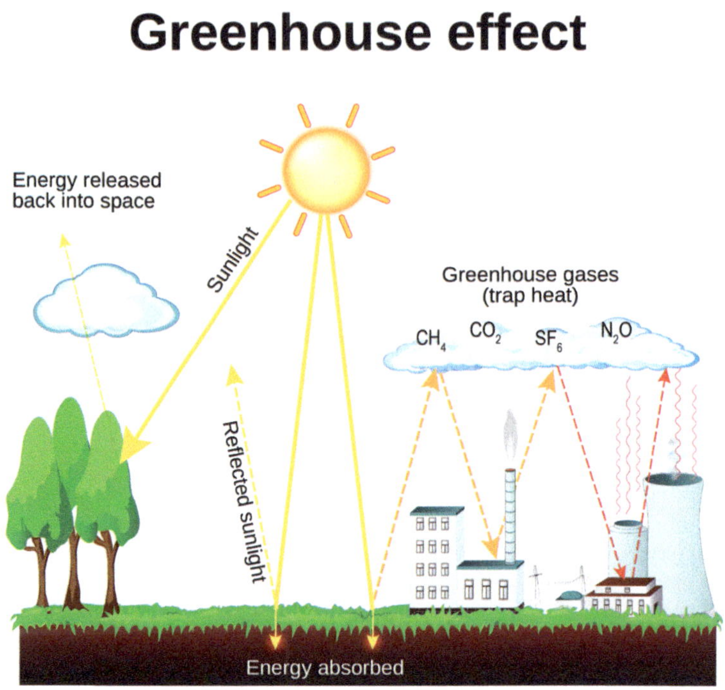

The greenhouse effect, shown here, traps gases in Earth's atmosphere, causing the climate to grow warmer.

The Greenhouse Effect

Certain gases hold heat in the earth's atmosphere. Like a greenhouse, they trap energy absorbed from the sun. These greenhouse gases are not all bad. Without them, Earth would be as cold as the moon. It would not support human life.

Carbon dioxide is a greenhouse gas that stores huge amounts of heat. It is emitted every time an animal breathes. Its levels in the atmosphere rise and fall naturally over time. In 1896, a Swedish chemist named Svante Arrhenius noted that powering the many factories that had arisen since the Industrial Revolution caused increased fuel consumption. Burning fossil fuels such as gasoline and coal increased carbon-dioxide emissions. Arrhenius predicted that these increased carbon emissions would cause Earth's temperatures to rise. They would also increase the amount of solar energy stored in the atmosphere. Arrhenius believed that the climate would grow warmer and severe weather would become more common as a result.

Humans' negative impact on the environment is nothing new. People have been burning coal and chopping down forests for thousands of years. Deforestation and poor farming practices are major producers of greenhouse gases. Modern innovations such as cars, airplanes, and electricity have greatly increased humans' carbon emissions as we burn huge amounts of fossil fuels.

The increases in carbon dioxide in the atmosphere correspond with the increase in human activities that produce it. Nearly all scientists agree that the changes happening in the twenty-first century are not natural. They say there is at least a 95 percent chance that human activity is a major cause of today's rapid global warming.

Another important greenhouse gas is methane. It is produced in especially large amounts by livestock. Livestock excrete methane as part of their natural digestion process. As people have begun to consume more meat, methane emissions have skyrocketed. Mining and manufacturing also produce large amounts of methane. Other chemicals such as coolants used in air conditioners further contribute to the greenhouse effect.

Extreme Weather

When the air temperature rises, heat waves are a natural effect. Heat waves cause droughts. Hot and dry conditions contribute to forest fires.

Polar ice caps are melting as a result of global warming. As they melt, the level of water in the oceans rises. Many coastal cities are built at or below sea level. Heavy rains leave them more vulnerable to flooding. Hurricanes combine heavy rains with storm surges that magnify high tides.

It might seem illogical that global warming can cause both droughts and floods, heat waves and blizzards. However, different regions experience different weather at different times. When the earth warms, extra moisture is trapped in the atmosphere through evaporation. This moisture can cause heavier precipitation in the form of rain and snow. Therefore, climate change can also cause more severe winter weather in certain areas. Monster snowstorms have become increasingly common on the eastern coast of the United States in the twenty-first century.

Researchers calculated that rising temperatures have resulted in nearly 2 trillion gallons (7.5 trillion liters) of additional water vapor in the atmosphere over the continental United States. They knew this evaporated water must fall eventually in the form of rain or snow. They predicted more intense rainstorms and flash flooding as the climate continued to warm. They were right. Very heavy rainstorms have increased by over 70 percent in the northeastern United States in recent decades. These storms often result in flash

High winds from hurricanes can flatten buildings.

floods. Flash floods can wash out bridges, roads, and dams. Dams are not always built to withstand such conditions, and their failure can lead to even more massive flooding. Because flash floods are difficult to predict, they are often deadly.

Some areas are used to receiving heavy snowfalls each winter. Blizzards can paralyze cities in more moderate climates that are unaccustomed to so much snow. Snow removal presents a huge challenge in these areas. Often there is no good place to put the snow, which slows plowing efforts. It may be bulldozed into surrounding waters where it worsens flooding when it melts. Moving large amounts of snow can also be a huge expense that local governments may not be prepared to pay for.

Hurricanes are megastorms that form over warm water when the air is warm and moist. They result in high winds and driving rains and can cause widespread flooding and damage. Scientific models show that climate change should make hurricanes occur more often. Evidence on hurricane frequency is mixed. However, recent hurricanes have shown a tendency to be stronger and to intensify more quickly than in the past. The US National Oceanic and Atmospheric Administration (NOAA) predicts that tropical cyclones will be on average up to 11 percent more intense

DID YOU KNOW?

Hurricanes, typhoons, and cyclones are different names for the same weather phenomenon. Their classification is determined by their location. Hurricanes occur in the Caribbean Sea and the Atlantic and Northeast Pacific Oceans. Typhoons affect the Northwest Pacific. Cyclones strike the South Pacific and Indian Oceans.

by the end of the twenty-first century because of climate change. The stronger a hurricane or cyclone is when it makes landfall, the more dangerous it will be. The more quickly a hurricane intensifies, the more difficult it is to give people ample warning to get out of its path.

The link between tornadoes and climate change is less clear. Tornadoes are caused by a combination of two elements. One is energy, which is produced by moist, warm air. The other is wind shear, or the change in the wind's direction and speed. Researchers believe that climate change should increase energy but reduce wind shear. Tornadoes have not become more frequent or destructive overall in recent years. However, as climate change becomes more pronounced, this pattern may change.

DID YOU KNOW?

Scientists estimate that rising seas caused about eighty-three thousand additional people in New Jersey and New York to experience flooding after Hurricane Sandy in 2012.

CLIMATE-CHANGE SKEPTICS

Some skeptics do not believe that climate change is occurring at all. Others acknowledge that Earth is warming but deny the role of human activity. They cite natural climate variances that the planet has experienced over the millennia.

Skeptics may point to individual events to question whether apparent trends are real. For instance, they may single out a particularly cold winter as evidence against a warming planet. While average temperatures are climbing, a single chilly winter does not undermine scientists' claims about overall warming. Looking at exceptions while ignoring major patterns is known as cherry-picking data.

The majority of scientists agree that climate change is accelerated by human activity. This has a major impact on weather systems and patterns throughout the world.

CHAPTER TWO

EXTREME WEATHER TODAY

It is very difficult to determine the exact cause of any outcome. Often multiple causes are combined. Looking at historical data and scientific models helps researchers estimate the effects of climate change on severe weather events. This growing field is called attribution science. Researchers look at patterns and odds. They try to evaluate causes separately to determine how much of an impact each had. A

Opposite: Climatologists study changes in weather conditions over time. This climatologist is working in Canada.

2012 California drought, for example, was primarily a result of natural climate conditions in that area. However, scientists calculated that it was about 15 to 20 percent more severe due to global warming.

Measuring Climate Change

Assessing changes in weather over time poses several challenges. While temperature is easily measured with a thermometer, official data collection didn't begin until the late nineteenth century. There is no way to look at historic highs and lows from many years ago. Measurement of rain and snow accumulation is similarly straightforward but lacking a strong historical record.

Today, there are several systems in place to measure extreme weather events. In 1971, the Saffir-Simpson scale was developed to measure hurricane intensity. It is based on wind speeds. Category 1 storms are the least intense; Category 5 hurricanes are the strongest. The Fujita scale measures tornado intensity and is also based on maximum wind speeds. Factors such as loss of life and property damage are other important measures of the destructiveness of a storm.

Climatologists calculate the odds that a flood will occur in any given location. Insurance companies and city planners use these data to determine where houses should be built. A very serious flood in a particular place may be expected every one hundred or five hundred years. As severe weather increases, these odds change. Three "once-in-five-hundred-years" floods occurred in Texas over the course of three years. Scientists will have to reassess these risks in light of climate change.

Death and Disease

One of the most obvious and terrible effects of extreme weather is loss of life. Many people have recurring nightmares about deadly earthquakes and tsunamis. The biggest natural disaster killer is much less dramatic: heat.

In 2003, a heat wave resulted in tens of thousands of deaths across Europe. By the time people realized how sick they were from heatstroke and dehydration, it was often too late to save them. Funeral directors had to rent space in refrigerated warehouses because they didn't have enough room for all the bodies. More people died that summer in France than in any year since World War II.

Young children and the elderly are most susceptible to heat-related illness. Even when there is another official cause of death such as heart disease, the temperature may be a major contributing factor. Therefore, the number of deaths due to heat waves is likely much larger than the official tallies.

In July 2010, the worst Russian heat wave in recorded history caused a severe drought. Massive forest fires spread quickly across the parched land. A haze of smoke and toxic chemicals seared people's lungs. Thousands lost their homes. More than half a million acres (202,342 hectares) were scorched. The suffocating heat and pollution killed over fifty-six thousand people.

DID YOU KNOW?

According to the World Health Organization, more than 150,000 people have died as a result of climate change in the past thirty years, and five million illnesses have resulted.

Along with more forest fires, heat waves can create dust storms in desert areas. In the summer of 2011, a dust cloud surrounded Phoenix, Arizona. Planes were grounded because pilots couldn't see. The sky turned nearly black. In Africa, giant dust clouds have formed that were thousands of miles long. These dust clouds spread viruses, bacteria, and illness-causing fungi.

The air also carries pollen, which flourishes over longer, warmer summers. The number of people suffering from allergies and asthma has risen sharply over the last forty years. Extreme temperatures and smoke from forest fires contribute to poor air quality and smog. Smog makes it even harder for people with asthma to breathe. It also contributes to other illnesses, such as lung cancer and dementia. It causes the heart's arteries to thicken more quickly and makes heart attacks and strokes more likely.

Hot, wet weather spawns other kinds of illnesses. Mosquitoes multiply rapidly in these conditions. They appear in thick swarms after floods and hurricanes. They carry and spread many deadly diseases, such as malaria and the Zika virus.

In the days and weeks following a hurricane, disease can be a major concern for several reasons. Muddy floodwaters teem with bacteria. Waste treatment plants are overwhelmed by heavy rainfall and may spill sewage into local waterways. In the aftermath of Hurricanes Katrina and Rita in the United States in 2005, hundreds of oil spills contaminated the nearby seas. After Hurricane Harvey in 2017, a chemical factory plant in Texas exploded due to

In 2017, flooding from Hurricane Harvey caused an explosion at a chemical plant in Crosby, Texas.

flooding, releasing massive amounts of toxins into the air.

Power outages after a major storm can affect access to clean running water. Refrigerated food quickly spoils. People stranded due to flooding may run out of basic necessities.

When hospitals flood, they have to evacuate even their sickest patients. Electrical outages may endanger people on life support or dialysis. Medical personnel may have trouble getting to the hospital to provide care.

Injuries from accidents also increase after a flood. Driving can be unsafe, with cars swept away by rushing waters. Downed power lines pose an electrocution hazard. Generators may be used to provide power when there is no electricity. They run on gasoline and can cause deadly carbon monoxide poisoning.

Delayed effects of flooding include toxic mold, which is especially dangerous to people with allergies and asthma. Floodwaters contaminated with chemicals and pesticides cause rashes and infections, including potentially deadly flesh-eating bacteria. Cancer risk may increase later in life from exposure to poisons. The costs of treating such illnesses are often not considered when calculating natural disaster damages.

Mental health issues are another common and costly result of natural disasters. A child whose house was swept away in a flood might experience depression or post-traumatic stress disorder for years to follow.

Poverty

An immediate effect of drought is lack of access to drinking water. If farmers don't have enough water to irrigate their fields, crops fail. Livestock die. Famine and death result.

A severe drought struck the midwestern United States in 2012. Corn crops failed. Water levels of the mighty Mississippi River reached historic lows. Boats could not cross over navigable areas that became too shallow. River traffic came to a near-halt in some spots. Cargo was stalled. Soybeans spoiled in storage while waiting to be transported. Prices soared. Farmers couldn't afford to feed their animals. Beef and milk prices rose as a result.

This photograph shows the Mississippi River at one of its lowest points during a drought in 2012.

When wells dry up and food is scarce, surviving residents must find new homes. Drought is one of the largest causes of refugee crises in impoverished areas of Africa and Asia. Destructive storms can also turn large numbers of people into homeless refugees. When people flee their homes, they often pack into refugee camps or city slums. Disease spreads easily in these hot, overcrowded living spaces.

A drought in Syria from 2006 to 2009 caused extreme starvation and worsened existing political unrest. Many residents fled their homes for cities. Unemployment and poverty in the cities exploded.

In neighboring Tunisia and Egypt, rising food prices caused riots. Conditions in the Middle East were ripe for the revolutions that swept the region in 2010. Desperation in the face of starvation often leads to war.

According to the Pentagon, climate change is one of the major threats to global security. In addition to causing unrest, natural disasters can affect military readiness. Soldiers are often called to help communities devastated by disasters. These duties divert resources and personnel from other important missions.

Developing countries experience the most impact from natural disasters related to climate change. Poor families are much more likely to live in dangerous conditions: on floodplains, steep hills, or homes that are not built to withstand storms. They don't have any other options. These people are the least able to weather the hardship that a flood or drought can bring.

If the world's temperature rises a few more degrees, researchers say mass starvation of hundreds of millions more people is possible.

Effects on Nature

When evacuating during a hurricane, people with animals face difficult choices. Many shelters will not accept pets. Some residents therefore refuse to leave their homes. This decision puts their lives in danger. If they do evacuate, they might have to leave their pets behind. Livestock present a similar dilemma. For farmers, they represent huge investments that can be total losses in a natural disaster. Thousands of animals starved or drowned after Hurricane Katrina in 2005.

Wildfires in Victoria, Australia, in 2009 were likely linked to global warming. These fires killed 173 people

Many pets were abandoned during evacuation from Hurricane Katrina in 2005.

and caused more than $1 billion in damage. They destroyed entire towns and huge swaths of land. A quarter of Victoria's wildlife, including koalas and kangaroos, were killed. Thousands of other animals lost their habitats. Many endangered species of plants and animals inched closer to extinction.

Wildfires lead to deforestation, which can cause massive land erosion. Erosion is also a natural effect of severe storms battering coastlines as sand and soil are washed into the ocean. Erosion in turn makes

Wildfires left this northern California landscape barren.

flood disasters more dangerous. It can lead to deadly mudslides and landslides.

Human Factors

Poor decisions by humans contribute to casualties from natural disasters. When people refuse to evacuate before a storm, they are risking not only their lives but also the lives of first responders who may be dispatched to save them.

The decision to leave home in the face of an oncoming storm can be very difficult. For residents who are ill or elderly, it may be impractical to pack up and leave. People with few financial resources may lack good options for transportation and lodging.

Many residents trying to evacuate New Orleans, Louisiana, before Hurricane Katrina had no way to get out of the city. Public transportation was overwhelmed, and large numbers of city dwellers had no cars. The roads were clogged with evacuees all traveling in the same direction.

Island inhabitants may be similarly stranded when a hurricane strikes. Many Caribbean islands are frequent hurricane targets.

Some people can't leave their homes before an impending storm because they have important jobs to do in their communities. They are doctors, nurses, firefighters, and police officers. They must stay behind to help others who cannot or will not leave. Others stay because they worry about lost wages from missing work. Unfortunately, many people choose to believe they can safely ride out a dangerous storm. Sometimes they're right. Sometimes they're not.

Houston's Growth

The city of Houston, Texas, was designed to withstand a once-in-one-hundred years storm. The likelihood of such a terrible storm occurring in any given year was supposed to be 1 percent. Reservoirs constructed in the 1940s had been designed to hold back Houston's floodwaters. Over the decades, channels were added to divert excess water to the nearby bayous.

As new roads were built and more people arrived, there was less room for storm water run-off in a heavy rainfall. Wetlands absorbed precipitation as well as wind force, helping to weaken strong storms. They provided an important protective buffer during a hurricane. However, Houston paved over 30 percent of its surrounding wetlands between 1992 and 2010. Fewer wetlands meant more flooding. The bayous could no longer soak up all the water that would fall in major storms. Meanwhile, storms kept getting bigger.

New roads were built to collect excess water if the storm drains failed. The drawback was that these roads would flood easily. If people needed to evacuate the city, the roads were impassable. Leaving Houston in an emergency would be difficult—or impossible.

There were opportunities to take more action and reduce damage or loss of life in the event of a major storm. In 1948, 1962, and again in 1993, the people of Houston had a chance to vote for new zoning laws. These laws would have prohibited building in low-lying areas that might flood. However, residents voted to let the city continue to grow unchecked. In 2008, city officials could have voted for building codes that would elevate new homes so they would better resist flooding. Instead, they chose to continue to allow homes to be more cheaply built but less safe.

Ultimately, Houston chose economic growth over increased safety. Houston was the largest major city in the United States without zoning laws. It was also one of the most successful economic growth stories in recent US history.

Then came a once-in-five-hundred-years storm in the spring of 2015. The city was hit hard by floods. People died. Almost unbelievably, the next year brought another once-in-five-hundred-years storm. More people died.

The federal government sent money for residents who lost their homes to rebuild. Many rebuilt within the once-in-five-hundred-years floodplain. They

Volunteers use boats to rescue trapped residents from a Houston, Texas, nursing home after Hurricane Harvey in 2017.

assumed another disaster couldn't possibly happen in their lifetimes. However, that changed in 2017 with the arrival of Hurricane Harvey.

Category 4 Hurricane Harvey bore down on the city in August 2017. The mayor, Sylvester Turner, knew that evacuating could be at least as deadly as staying. City officials urged people to stay put.

Then the floods came. Fifty inches (127 centimeters) of rain fell. Water rushed into low-lying homes. Boats had to go door-to-door to rescue trapped citizens. Residents of a nursing home built on the floodplain were left sitting in water up to their waists. Dozens of people in the Houston area were killed. Better planning, engineers said, could have saved lives and prevented millions of dollars of damage from the storm.

DID YOU KNOW?

The economic toll of weather disasters in the United States since 1980 is $1.2 trillion and counting. This value does not account for other effects such as loss of life, damage to the ecosystem, and physical and mental health costs over many years.

Draining the Florida Swamp

With the world's population rapidly growing, many people are settling in places that were once considered uninhabitable. Southern Florida is one example.

The first American visitors to southern Florida found a swampy, sweltering, mosquito-infested wasteland. The government built a makeshift system of dikes and canals to dry out the land and make it livable. Glitzy beach towns such as Miami and Fort Lauderdale began to spring up. In 1926, a local newspaper assured residents that the risk of walking across the street was larger than that of a deadly hurricane. Days later, Miami was flattened by a Category 4 storm. Another followed two years later. At the time, it was the second-deadliest US natural disaster. The south Florida real estate boom ended abruptly.

The US Army Corps of Engineers was then called in to perform a miracle. They built a sophisticated system to drain much of the swampy Florida Everglades. Levees, canals, and pumps turned marshes into dry land, which was paved over with highways.

THE FLOODING OF THE NILE

In northern Africa lies the Nile River. Even in ancient times, its banks provided useable land in the desert terrain. No matter how much it rained, the Nile overflowed each summer. This pattern was caused by annual rains at the distant headwaters of the river.

The ancient Egyptians considered the flooding a blessing from the gods. The water overflowed into the surrounding plains and fields. It brought with it rich, black, fertile soil. The silt from the river also formed natural levees. The floodwaters provided reliable irrigation for crops. This bounty allowed the civilization to flourish.

Around 3000 BCE, the Egyptians began to try to extend the river's reach. They built canals, channels, and levees to irrigate larger areas. Rather than holding the floodwaters back, they used these structures to improve their quality of life.

The Everglades is a large area of swampy wetlands in south Florida.

Millions of people traveled on these roads to settle in south Florida.

Today, this area is highly susceptible to floods, hurricanes, and fires. Hurricane Irma's path in September 2017 barely missed many of the most vulnerable and populous areas. However, as the seas rise, it's almost inevitable that many of the homes in these places will eventually be reclaimed by nature.

California Droughts and Floods

Unlike soggy south Florida, much of California is plagued by drought. Los Angeles, the second-most populous city in the United States, has hot, dry summers. Every few years, the area experiences very wet winters caused by a Pacific weather pattern known as El Niño. Climate change has worsened the periodic droughts and the floods in the region. Both endanger the city's drinking water.

Los Angeles was once a relatively small town. As its population began to grow in the early twentieth century, the water supply could not keep pace. The Los Angeles River slowed to a trickle in some places. The city desperately needed to find another water source.

Reservoirs were built to carry water from the upstate Owens River to the residents of thirsty Los Angeles. Much of the Owens Valley soon became a dusty wasteland. Residents abandoned villages and farms. Meanwhile, the city's steady water supply allowed it to grow rapidly. As more people came, the Los Angeles aqueduct, a main way of bringing water into the city, could no longer supply the ever-increasing population. More complex systems were constructed to carry water from the northern part of the state, where most precipitation fell, to the drier central and southern parts where more people lived.

California endured a five-year drought that ended in the winter of 2016 with the wettest season since recordkeeping began. During the drought, thousands of wells ran dry, forcing people to abandon their homes.

Lake Oroville is one of the most important man-made bodies of water in the state. It supplies drinking water to twenty-five million people in Los Angeles and irrigates farms throughout the Central Valley. In 2015, the water level in Lake Oroville was lower than it had been in forty years. The next year, the lake was inundated with floodwaters. These put massive pressure on an aging dam that threatened to give way. Nearly two hundred thousand residents had to be evacuated. Engineers barely averted a massive, deadly flood that would have endangered the water supply of millions of people.

Parts of California's Lake Oroville (*top*) became nearly dry (*bottom*) during a drought in 2014.

ROMAN AQUEDUCTS

Ancient cities formed on the banks of rivers. The water was their lifeblood. It transported supplies by boat. It helped crops grow. Most important, it gave easy access to the drinking water people needed for survival.

The Romans' invention of aqueducts helped the Roman Empire expand and flourish. These elaborate systems carried water from mountains and underground springs to distant sites. The aqueducts followed the natural slope of the land and used gravity to keep the water moving.

Aqueducts allowed new cities to be built on dry plains. Easy access to clean water helped reduce disease. Public baths became popular gathering places.

The idea to divert natural water supplies had huge effects—both positive and negative—on future civilizations. Roman aqueducts serve as a model for many major water systems today.

CHAPTER THREE

TWENTY-FIRST-CENTURY INNOVATIONS

In the twenty-first century, many people and businesses are taking steps and developing technologies that can help reduce carbon emissions, which are key contributors to global warming and lead to increases in extreme weather events. In a world threatened by changes in climate and weather patterns, some of these new inventions and actions being taken also attempt to curb extreme weather events themselves.

Opposite: In 2013, Google displayed Project Loon balloons in New Zealand. They can provide internet service in areas that don't already have access.

A Local Approach

Natural disasters are extremely costly and becoming much more common. To reduce the costs of natural disasters and improve public safety, effective solutions are needed and actions should be taken. Local governments are motivated to make changes that immediately improve the lives of their citizens. For example, many cities in coastal areas in the United States have evacuation plans and routes that help move people from one part of a state or region to another, toward higher ground and away from danger. Almost all cities have relief centers. These are places where people can go when extreme weather hits, if they don't have a secure space within their home. Many cities also have emergency supplies and workers on hand to help rescue people after an extreme weather event.

Low-Tech Approaches to Rising Seas and Temperatures

Today, more people than ever are living in cities, and many large cities are situated in coastal areas. Coastal

BUSINESSES AND CLIMATE CHANGE

Today, some businesses have adopted policies to reduce their carbon footprint. Others have embraced the minority of scientific doubts that humans cause climate change.

Governing bodies around the world have made several suggestions on how companies can lower carbon emissions. These changes include using more recyclable packaging and creating more environmentally friendly products.

The changes manufacturers must make to reach safer carbon figures may have a significant impact on profits. Without public demand or government mandates, many companies lack the motivation or the will to make these changes.

Taking measures to fight climate change will be extraordinarily expensive in the short term. However, reducing carbon emissions could help save the planet and all life on Earth.

New York Mets players participated in the 2010 Cool the Roof project in Brooklyn. They covered the roof in reflective paint.

areas often face increased threats of floods from rising sea levels. They also experience many extreme weather events, such as hurricanes, tsunamis, and storm surges.

One extreme weather event that is being discussed more as temperatures increase is heat waves. Tall buildings and asphalt absorb heat. Therefore, city temperatures on average are 5 to 10°F (2.7 to 5.5°C) higher than those in surrounding areas. Many people in cities have come to use air-conditioning systems to fight the soaring temperatures. However, the use of air conditioning contributes to carbon emissions. New technologies are being developed to combat increasing temperatures and a reliance on air conditioning. For example, in 2009, the mayor of New York City, Michael Bloomberg, launched a Cool Roofs campaign. Businesses and landlords across the city were urged to paint their roofs lighter colors or make them reflective. This would, in turn, make the roofs cooler. Laws required new roofs to meet these guidelines. Businesses and landlords who took up the challenge would experience a quick return on their investments, as they could spend less money on cooling costs.

Another means to combat the heat is the opening of cooling centers. These facilities are used by people who cannot afford to cool their homes with air conditioning. In extreme heat, lack of air conditioning presents a major health hazard and can lead to people getting heatstroke. These cooling centers provide respite from the heat for people who lack air conditioning at home. Also, gathering people in one place to stay cool requires less energy consumption than cooling each individual home separately.

Restoring Utility Service

Electrical outages have become more common in the twenty-first century due to extreme weather events. Heat waves can cause blackouts due to large electricity demands from air-conditioning use, and

Electrical outages occur when lightning and high winds damage power lines.

fierce lightning strikes can take out transformers, which generate power to entire regions. Overheated equipment may stop working or even catch fire. Lasting damage and long power outages may result.

Power companies have begun to use more wind and solar energy to supply electricity to growing populations. A network of generators, stations, and cables is known as a power grid. Today's grid managers use complex computer algorithms to deal with varying supplies and demands. They have developed technology to store power for later use and can even share it with other states when they have more than they can store. Smart meters allow power companies to remotely adjust customers' thermostats to help conserve power. Smart grid technology can distribute electricity more efficiently. Fewer spikes in usage mean that the system is less likely to be overwhelmed and shut down too.

Blackouts are also common after hurricanes and other major storms. Strong winds, thick ice, and snowstorms can snap power lines and topple electrical poles. In parts of the world, people bury electrical cables underground. This can prevent many power outages, but doing so is expensive. In the United States, some power companies have begun replacing wooden poles with sturdier concrete ones. Another simple but important preventive measure against blackouts is maintenance of trees near vulnerable electrical wires.

Flooding at power stations can short-circuit equipment and cause costly damage. Digital flood monitors allow power stations to shut down before they are swamped with water, preventing damage to electrical equipment that would be time-consuming to repair. While customers still experience a service interruption, power is restored much more quickly after a storm.

Restoring power quickly after a storm or blackout is a matter of life and death. Therefore, identifying the sources of electrical outages is important. However, sometimes restoring power can take many hours, even days, as workers fan out across large areas after a storm to assess damage. In more developed countries, advanced sensors and electrical meters can help technicians instantly locate problems and restore power more quickly. Designing smaller power grids, called microgrids, can allow undamaged areas to detach from the system and function independently.

Restoring communication after a storm is also vital. Hurricanes can destroy cell phone towers, depriving residents of contact with their loved ones around the world. One company that is changing how people communicate after an extreme weather event is Google's Project Loon. This organization sends giant balloons about 11 miles (17.7 km) into the air after a storm. Each balloon carries solar technology that relays phone signals from the ground to the air. They create an in-air wireless communications network. In 2017, Project Loon deployed in major areas in Puerto Rico after it suffered extensive damage and an island-wide electrical outage from Hurricane Maria. This allowed people who had been out of touch with loved ones for weeks to finally get through to family members and request help or update them on their condition.

Planning

Improving disaster preparedness is extremely important as climate change makes major weather events more frequent and more severe. Today's technology allows instant access to information. In the twenty-first century, warning people of impending danger is easier than ever.

First responders have begun to harness the power of social media to locate people who need to be rescued, and software engineers have created apps to help people locate available food, water, and gas during storm-related shortages. Some examples of helpful apps in emergency situations are the Transitional Sheltering Assistance Hotel Locator app, which helps people who need it find shelter; the Life360 app, which helps family members connect to loved ones who may be in an extreme weather emergency; and the Gas Buddy app, which tells people where the most affordable gas stations are—useful in an evacuation situation.

Likewise, researchers have begun to crunch data to help determine when and whether evacuation is advisable and, if so, what the best routes are. Artificial intelligence uses satellites and drones to map disaster zones in real time. First responders can then allocate resources according to current conditions on the ground. Snow removal after major storms is also more efficient thanks to GPS-based technology that has become available in the early twenty-first century.

Architects and civil engineers are planning new construction projects that will be less vulnerable to the effects of severe weather. Stricter building codes and zoning measures that prohibit building in floodplains reduce the loss of life and property in natural disasters. Avoiding new construction in areas that are too arid or marshy is another common-sense approach. City planners also help reduce the risks of severe weather in the long term by adopting measures that decrease carbon emissions. They are designing neighborhoods to be more walkable and more accessible for pedestrians. Quality public transportation, bike-sharing initiatives, and tolls on city roadways also reduce fuel emissions.

Weather Forecasting

Two key solutions to preparing for weather events are improved computer technology and twenty-four-hour newsrooms. Having almost instant access to real-time weather radar and keeping up to date with weather forecasters as extreme weather approaches helps people prepare and survive extreme weather in their area. More localized forecasts help people better understand what type of weather to expect in their individual neighborhoods. More accurate tracking models for hurricanes give a clearer idea of the paths storms will take and help determine when and whether people should evacuate. If people know they can trust the weather forecast, they are much more likely to evacuate when their lives depend on it.

Having these tools in place proved extremely useful in 2017. In September of that year, residents in southern Florida were told to evacuate areas where

DID YOU KNOW?

According to a January 2018 article on cnn.com, the year 2017 was the costliest year for weather- and climate-related events in the United States. In total, sixteen weather- and climate-related disasters cost over one billion dollars in damage. The majority of damage-producing events were hurricanes, floods, droughts, and wildfires.

a potentially catastrophic Hurricane Irma could make landfall. Millions of people took these warnings seriously and moved to higher ground.

Weather satellites have also become important in predicting weather events. The first of three new US weather satellites was launched in 2016 to provide more precise and current data about atmospheric conditions. That same year, NOAA also began developing a new model that it plans to roll out by 2019. Today, forecasts typically give weather information eight to ten days in advance. NOAA aims to provide fourteen-day predictions in the near future. Meteorologists also hope to provide more warnings for extreme weather events, up to three or four weeks ahead of time.

Flood Control Technology

Flood control is increasingly important in the twenty-first century, especially as flooding becomes more frequent following extreme weather events. However, flooding is not just a twenty-first-century problem. It also affected ancient peoples, who had their own ways of dealing with flooding.

Humans used to be nomads. They wandered from place to place and did not build permanent settlements. Eventually, that lifestyle changed. People began to settle in communities. As towns and cities were built on riverbanks, people could no longer pick up and move when the river flooded. They developed a variety of ways to control the water and protect their homes. Two methods were canals and channels. Both divert water from a river's natural path. These

The Hoover Dam supplies hydroelectric power to people living in Nevada, Arizona, and California.

new paths relieved pressure on certain parts of the river. They could also be used to provide irrigation to new areas.

The ancient Egyptians, Sumerians, and Romans became known for their flood management and prevention methods. The Egyptians created the Sadd el-Kafara dam around 2700 BCE to stem the Nile River, although it didn't last long. The Sumerians built Nimrod's dam, as well as canals and levees. The Romans had similar flood-prevention techniques. Their dams were unique, though. They were formed using concrete, which had not been used by other civilizations.

Dams, levees, and dikes were used to stem or trap water. A dam is built across a river to trap water and prevent flooding. Levees are used to protect areas that flood occasionally. Dikes are walls built to reclaim land from areas that would otherwise be underwater.

Modern flood-control measures often use a complex combination of levees, dams, canals, pumps, and other related technology. Some of these systems also generate hydroelectric power, produced by turbines that are spun by rushing rivers. They collect

A dam's spillway is designed to control the release of water when reservoir levels grow too high.

water in lakes called reservoirs that provide drinking water to communities.

Most human efforts to control and prevent flooding have produced mixed results. Under normal conditions, a river's banks prevent its water from overflowing. Heavy rain will occasionally cause extreme flooding. Levees and dikes protect communities from these floods. However, they cause tension in other areas of the river that are forced to hold more water. Sometimes, levees are poorly built. Sometimes, they deteriorate. Sometimes, they are simply not big enough to do the job, especially as severe flooding becomes more common due to climate change.

When a reservoir overfills from flooding, the dam's spillway is designed to release the excess water. In order to protect a community downstream, engineers may have to make a difficult decision to flood the area surrounding the spillway. Though this release is done in controlled fashion, it can cause massive damage to nearby homes.

The Mississippi River

Today, the Mississippi River provides fertile farmland and an avenue of transportation for crops and goods. About 41 percent of the continental United States's land drains into the river's basin.

In order to protect the cities and towns along the Mississippi's banks, people have constructed over 3,500 miles (5,632 km) of levees. These levees divert silt as well as water. However, there are negative

FREDERICK, MARYLAND

A 1976 rainstorm in Maryland turned calm Carroll Creek into a raging flood that roared into the city of Frederick's downtown business district. A wall of water and mud hit blocks of restaurants and stores. The flood caused up to $25 million in damages. The city's economy took years to rebound.

City planners knew that Frederick would have a very difficult time recovering from another disaster of this magnitude. They needed to take measures to prevent future flooding. The most practical solution was a drainage system that would funnel drainage from the creek into the Monocacy River. This plan would be effective but unattractive.

Architects looked at a variety of other options and developed a novel solution that pieced together the best parts of several ideas. They built a man-made creek over top of the flood control system. The elevated creek was lined with pedestrian walkways, bike paths, shops, and restaurants. Water lilies controlled algae, and a system of pumps and valves kept the creek's current moving. The surrounding park provided an appealing environment as well as a buffer to absorb storm water.

The city's proposal to invest in this expensive plan was controversial because of the costs. Today, it has attracted over $1.5 million in private investments and in 2018 was estimated to add 1,500 jobs to the local economy. Frederick is reaping the economic benefits of the revitalized downtown.

effects to the diversions. For instance, the silt clogs dams, reducing their ability to produce power. Prior to being diverted by levees, silt used to flow down the river to low-lying New Orleans. There it would collect and build up the land, protecting the city to some degree from the rising seas. Now, without this silt, wetlands in the area are washing away.

Levees are designed to keep the river from spreading beyond its banks. If the water can't spread, it grows higher. With very heavy precipitation, levees can actually make flooding worse.

Today, floodwaters from severe storms routinely top many of the levees along the Mississippi River. Residents must resort to using stacked sandbags as temporary levees that may or may not save their homes. Companies have even begun manufacturing inflatable floodwalls that can be deployed to protect individual homeowners' properties when the engineer-built floodwalls fail.

The US Army Corps of Engineers determined that the main cause of the floods that inundated New Orleans after Hurricane Katrina in 2005 was failure of the levees. In 2012, they finished rebuilding the flood control system surrounding New Orleans. The levees are now 12 feet (3.7 m) taller. They are also stronger. During a flood, a gate the size of a football field swings shut. Eleven enormous engines start pumping water out of the surrounding floodplain. The project, which cost over $1 billion to complete, includes an enormous network of floodgates, pumps, and canals. The city is now deemed safe from a once-in-one-hundred-years storm. However, another event like Hurricane Katrina would be devastating even to the new and improved system.

Ultimately, scientists say that rebuilding the wetlands along the river is the only way to give the water a safe place to go. Engineering solutions work best when used together with ecosystem restoration.

Water Access

As droughts, another extreme weather event, become more frequent due to climate change, capturing fallen rainwater will be more and more important. Los Angeles is exploring ways to recycle and reclaim it. The local government has invested in a water purification project that is slated to add 9.7 million gallons (37 billion liters) of recycled water to the supply by 2022. The city is also planning to build a huge wetlands park. It will catch storm water and pump it into underground wells.

Another Southern California city, San Diego, is trying a different approach to finding new water sources. Authorities have built one of the world's largest desalination plants to extract salt from seawater and turn it into drinkable freshwater. However, environmentalists are concerned about potential effects the plant might have on marine life and the ecosystem, which are largely unknown.

Cloud Seeding

For centuries, farmers have prayed for rain during times of drought. Sometimes, the rain would come. Other times, the drought would continue for weeks or months. New technologies allowed for research into weather manipulation in the twentieth century. In the 1940s, scientists began developing a technique called cloud seeding. Initially, they hoped that cloud seeding might allow them to douse forest fires with man-made rain or whip up white Christmases on demand.

Cloud-seeding projects are being developed to try to control precipitation. This airplane is cloud seeding above North Dakota.

Airplanes or flares carrying this dry ice "seed" would spray it into clouds in an attempt to get the clouds to produce more rain or snow.

While the technology did not deliver such spectacular results initially, interest in cloud seeding

DID YOU KNOW?

According to *Business Insider*, China reported an increase in precipitation through cloud seeding of 19.4 trillion cubic feet (55 billion cubic meters) between 2006 and 2016.

has increased in the twenty-first century—especially as extreme weather events grow more common due to climate change. China used cloud seeding to try to control the weather at the opening ceremonies of the 2008 Beijing Olympics. In 2017, they spent $168 million on new cloud-seeding technologies. Canada has used cloud seeding to try to reduce the impact of severe hailstorms. Ten states in the United States have been working on cloud-seeding projects, and power companies and private investors continue to do new research. While the data on its success are mixed, scientists say that cloud seeding could increase precipitation in a given weather event by up to 5 to 15 percent.

Hurricane Fighting

There is as much heat energy in a hurricane as there is in a nuclear bomb. Rather than resigning themselves to dealing with hurricanes' effects, scientists have experimented with ways of stopping them in their tracks by dissipating the accumulating energy.

In 1947, a Nobel Prize–winning chemist proposed fighting hurricanes with a form of cloud seeding. At his direction, a bomber plane dropped 180 pounds (81.6 kilograms) of shaved dry ice into storm clouds gathering over Florida. The ice did nothing to slow down the storm, unfortunately.

Other modern-day scientists and engineers are working on different methods. One such person is engineer Stephen Salter. Because warm waters feed hurricanes, in 2009, Salter designed a giant tube called the Salter Sink to try to cool the ocean. A pump powered by waves would circulate surface water into the cooler depths of the sea. He hypothesized that thousands of these machines working together might be able to reduce a hurricane's strength. Despite having Microsoft founder Bill Gates as an investor, as of 2017, Salter's idea remains untested.

However, Salter's ideas continued. In 2012, he developed another hurricane-fighting approach. He proposed using a fleet of unmanned boats to spray water droplets into the sky. The extra water surrounding the clouds in the sky would make them more reflective. This process, called cloud brightening, would theoretically divert some of the sun's rays from the water. In theory, it could reduce evaporation while simultaneously cooling the water below. Unfortunately, creating a prototype to test Salter's proposal would cost approximately $40 million. No one has yet volunteered to finance the project. However, scientists continue to experiment with the potential effects of cloud brightening.

Many climate researchers believe that ideas such as Salter's are totally impractical when it comes to fighting hurricanes. In order to be effective, they would have to be deployed at exactly the right time in the right place. They would have to cool huge areas of the sea very quickly. Many storms never form into full-blown hurricanes, so tremendous resources could be wasted on a scheme that would never work.

Climate Change Costs

The costs of dealing with more frequent and intense storms as a result of climate change are staggering. A 2017 study found that for every degree Fahrenheit of climate change, the United States is projected to lose 0.7 percent of its economic productivity from property and wage loss due to rising sea levels and natural disasters.

However, financial costs are only a small part of the story. Researchers predict that because of climate change, the death rate will rise. The crime rate will rise. Hunger will increase as food production decreases. Disease will spread. The stress of daily living will soar. While the short-term costs associated with fighting climate change may seem daunting, the long-term costs of failing to fight it are much larger.

According to Katharine Hayhoe, director of the Climate Science Center at Texas Tech University, the most dangerous belief Americans have about climate change isn't that it doesn't exist or that it isn't man-made. It's that it doesn't affect each and every one of us.

Climate change deniers argue that there is nothing humans can or should do to stop the onslaught of global warming. However, scientific evidence shows that many of its effects can be prevented, stopped, or even reversed. The longer we wait, the harder it will be. At some point, it may become impossible.

GLOSSARY

basin Land that is drained by a river system.

bayous Marshy outlets of a body of water.

canal A man-made passage that redirects water from an ocean or a river.

climatologist A scientist who studies climate.

cloud seeding A process by which a chemical such as dry ice is sent into the sky in the hope of creating precipitation in a desired area.

ecosystem A group of organisms that interact in a given environment.

erosion The gradual destruction or wearing away of land by nature.

flash flood Quick flooding, usually from a severe thunderstorm.

floodplain Low-lying ground near a river that frequently floods.

hydroelectric power Electricity generated by running water.

irrigate To provide water, especially for crops.

landfall When a hurricane makes contact with land.

levee A wall that holds back river water.

meteorologists Weather forecasters.

reservoir A large body of water that is used as a water supply.

spillway Part of a dam, designed to release water as needed.

FURTHER INFORMATION

Books

Cosgrove, Brian. *Weather: Discover the World's Weather—from Heat Waves and Droughts to Blizzards and Floods.* London, UK: Dorling Kindersley Limited, 2016.

Furgang, Kathy. *National Geographic Kids Everything Weather: Facts, Photos, and Fun That Will Blow You Away.* Washington, DC: National Geographic Society, 2012.

Kostigen, Thomas. *Extreme Weather: Surviving Tornadoes, Sandstorms, Hailstorms, Blizzards, Hurricanes, and More!* Washington, DC: National Geographic, 2014.

Snedeker, Joseph. *The Everything Kids' Weather Book: From Tornadoes to Snowstorms, Puzzles, Games, and Facts*

That Make Weather for Kids Fun! Avon, MA: Adams Media, 2012.

Streissguth, Thomas, and Michael E. Mann. *Extreme Weather.* Detroit, MI: Greenhaven Press/Gale, Cengage Learning, 2011.

Websites

Kids Ahead: Extreme Weather

http://kidsahead.com/subjects/5-extreme-weather

This site includes articles and games to engage kids in the subject of extreme weather.

The National Severe Storms Laboratory

http://www.nssl.noaa.gov/education/svrwx101

Severe Weather 101 gives essential information about weather phenomena such as floods, wind, and winter weather.

Weather Wiz Kids

http://www.weatherwizkids.com

This meteorologist's website is designed to get kids interested in weather.

SELECTED BIBLIOGRAPHY

Allen, Greg. "Army Corps Makes Tough Calls With Floods." NPR, May 9, 2011. http://www.npr.org/2011/05/09/136056393/army-corps-makes-tough-calls-with-floods.

Aschwandan, Christie. "We May Never Know How Well Cloud Seeding Works." FiveThirtyEight, March 6, 2016. https://fivethirtyeight.com/features/we-may-never-know-how-well-cloud-seeding-works.

Boburg, Shawn, and Beth Reinhard. "How Houston's 'Wild West' Growth May Have Contributed to Devastating Flooding." *Washington Post*, August 29, 2017. https://www.washingtonpost.com/graphics/2017/investigations/harvey-urban-planning/?utm_term=.7dfb7d732bf9.

Burnett, John. "Billions Spent On Flood Barriers, But New Orleans Still A 'Fishbowl'." NPR, August 28, 2015. http://www.npr.org/2015/08/28/432059261/billions-spent-on-flood-barriers-but-new-orleans-still-a-fishbowl.

"Climate Change Evidence: How Do We Know?" NASA, August 10, 2017. https://climate.nasa.gov/evidence.

Drash, Wayne. "Yes, Climate Change Made Harvey and Irma Worse." CNN, September 19, 2017. http://www.cnn.com/2017/09/15/us/climate-change-hurricanes-harvey-and-irma/index.html.

Gillis, Justin. "California Drought Is Made Worse by Global Warming, Scientists Say." *New York Times*, August 20, 2015. https://www.nytimes.com/2015/08/21/science/climate-change-intensifies-california-drought-scientists-say.html.

———. "Climate Change Is Complex. We've Got Answers to Your Questions." *New York Times*, September 19, 2017. https://www.nytimes.com/interactive/2017/climate/what-is-climate-change.html.

———. "Looks Like Rain Again. And Again." *New York Times*, May 12, 2014. https://www.nytimes.com/2014/05/13/science/looks-like-rain-again-and-again.html?_r=1.

Greenemeier, Larry. "How Will the Smart Grid Handle Heat Waves?" *Scientific American*, July 27, 2010. https://www.scientificamerican.com/article/smart-grid-heat-wave.

Harris, Mark. "Alphabet Closer to Using Balloons for Telecom in Puerto Rico." *Wired*, October 7, 2017. https://www.wired.com/story/google-closer-to-using-balloons-for-telecom-in-puerto-rico.

Hertsgaard, Mark. "How to Feed the World After Climate Change." *Slate*, October 26, 2017.

http://www.slate.com/articles/technology/future_tense/2012/04/heat_resistant_seeds_ecological_agriculture_growing_food_after_climate_change_.html.

Jefferson, Anne. "Levees and the Illusion of Flood Control [Explainer]." Scientific American Blog Network, May 20, 2011. https://blogs.scientificamerican.com/guest-blog/levees-and-the-illusion-of-flood-control-explainer.

Kunzig, Robert. "Tornadoes and Global Warming: Is There a Connection?" *National Geographic*, May 22, 2013. http://news.nationalgeographic.com/news/2013/05/130522-tornado-climate-change-oklahoma-science-global-warming.

Little, Amanda. "Weather on Demand: Making It Rain Is Now a Global Business." Bloomberg.com, October 28, 2015. https://www.bloomberg.com/features/2015-cloud-seeding-india.

Marsa, Linda. *Fevered: Why a Hotter Planet Will Hurt Our Health and How We Can Save Ourselves*. Emmaus, PA: Rodale, 2013.

Miller, Brandon. "US Shatters Record for Disaster Costs in 2017." CNN, January 8, 2013. http://www.cnn.com/2018/01/08/us/2017-costliest-disasters/index.html.

Moore, Tom. "NOAA Will Develop and Implement a New Global Forecast Model in Three Years." Weather Channel, July 28, 2016. https://weather.com/news/news/improved-global-extended-weather-forecast-model-accuracy-gfs.

NOAA. "Global Warming and Hurricanes." Geophysical Fluid Dynamics Laboratory. August 30, 2017. https://www.gfdl.noaa.gov/global-warming-and-hurricanes.

Pope, Carl, and Michael Bloomberg. *Climate of Hope: How Cities, Businesses, and Citizens Can save the Planet*. New York: St. Martin's Press, 2017.

Popovich, Nadja. "From Heat Waves to Hurricanes: What We Know About Extreme Weather and Climate Change." *New York Times*, September 15, 2017. https://www.nytimes.com/interactive/2017/09/15/climate/does-climate-change-cause-hurricanes-drought.html.

Sakoh, Mayumi. "Unnatural Disasters." International Fund for Animal Welfare, 2009. http://www.ifaw.org/sites/default/files/unnatural_disasters_0.pdf.

"Watering Ancient Rome." PBS, February 22, 2000. http://www.pbs.org/wgbh/nova/ancient/roman-aqueducts.html.

Zimmermann, Kim Ann. "Hurricanes, Typhoons and Cyclones: Storms of Many Names." LiveScience, September 5, 2017. https://www.livescience.com/22177-hurricanes-typhoons-cyclones.html.

INDEX

Page numbers in **boldface** are illustrations.

ABOUT THE AUTHOR

Jeanne Marie Ford is an Emmy-winning television scriptwriter and holds an MFA in Writing for Children from Vermont College. She has written numerous children's books on a variety of subjects, including Hurricane Katrina. She also teaches college English. She lives in Maryland with her husband and two children.